Joseph Edwards

How a person threatened or afflicted with Bright's disease

ought to live

Joseph Edwards

How a person threatened or afflicted with Bright's disease ought to live

ISBN/EAN: 9783337197582

Printed in Europe, USA, Canada, Australia, Japan

Cover: Foto ©berggeist007 / pixelio.de

More available books at **www.hansebooks.com**

HOW A PERSON THREATENED

OR AFFLICTED

WITH BRIGHT'S DISEASE

OUGHT TO LIVE.

By JOSEPH F. EDWARDS, M. D.

PHILADELPHIA:

PRESLEY BLAKISTON,

1012 WALNUT STREET.

1881.

Press of WM. F. FELL & CO.
1220-1234 Sansom St., Phila.

*With the hope that it may convince
him that the seeds which he
planted have not fallen
on barren ground, this first public fruit of his
teaching is affectionately and respectfully
dedicated by the author to his
friend and former
preceptor,*

Dr. Walter F. Atlee.

CONTENTS.

·PREFACE.

The necessity for this little work has been suggested to the author by many considerations: among them—

1. The great prevalence and fatality of Bright's Disease of the Kidneys.

2. Its insidiousness; the disease in many cases becoming irrevocably fixed and far advanced before it develops symptoms of sufficient importance to induce the patient to seek professional advice.

3. The fact that in many instances a person with a well marked case can, by leading a proper life, live in comfort and comparatively good health for many years.

4. That very few diseases are so liable to be aggravated by neglect of hygienic rules.

5. That being, as a rule, a protracted disease and one in which but little discomfort is experi-

enced until toward the end, the advice of the physician is apt to be neglected.

These and other considerations have induced the author to give, in this small work, in familiar, non-professional and easily understood language, a little history of this disease and some rules of life the faithful observance of which will insure to the sufferer *from this disease* the longest lease of life and the greatest amount of health of which he is capable.   If repetition may render the reading tiresome, I will say that all the precepts which are repeated are of very great importance, and that this very repetition will serve to impress them the more firmly on the mind.

JOSEPH F. EDWARDS.

Lansdowne, Del. Co., Pa.,
October, 1880.

HOW A PERSON THREATENED

OR AFFLICTED WITH

BRIGHT'S DISEASE OUGHT TO LIVE.

PART FIRST.

GENERAL REMARKS.

There are several diseases of the kidney, differing in the appearances which the organ presents after death, but so similar in their symptoms, progress and result, that they are generally grouped, and, indeed, are only known to the non-professional public, under the generic term of "Bright's Disease." This disease has received its name from Dr. Richard Bright, of London, who, in the early part of the present century (1828), added much to the previously limited knowledge of this affection. This disease no doubt existed prior to the time of Bright, and has probably existed for many years. In the older records of death we find many cases set down as occurring from causes *unknown*, *suddenly*, or from some cause which can be attributed to Bright's Disease; therefore, it is but fair to infer that some of these deaths may have been due to the disease under consideration; but the means of recognizing its existence being but little understood before the time of Bright, and certain cases not presenting the symptoms of any other disease, have been set down as *unknown*.

Many people, and even some physicians, have a bad habit of confounding a symptom of disease with the disease itself. For example, you often hear of a person dying from dropsy; now dropsy is not a disease in itself, but only a symptom of disease of some organ. Thus a contracted liver will press on the blood vessels passing through it, and interfere with the free passage of the blood, while the heart, acting from behind, will force the blood against this obstruction; now you can readily understand how this will cause a damming up, a congestion of the blood, and some of its water (seventy-nine per cent. of blood is water) will ooze out through the walls of the blood vessels into the surrounding tissue, and we have dropsy; but you see the disease is really seated in the liver, while the dropsy is only a symptom. Again, you will often hear of a person dying of convulsions. Here, too, the convulsion is only a symptom of the disease which causes it, and not really a disease in itself. Thus, in many cases of Bright's Disease the urea which is retained in the blood, being carried through the circulation to the brain, will irritate this organ and cause convulsions, and the patient may die in one of them. Here we really have death produced by Bright's Disease of the kidney, but the convulsion being the most prominent symptom, and the other manifestations being comparatively slight, they are lost sight of, and we have the effect given as the cause, and death attributed to the convulsion. Now, when we consider that this error is very common at the present day, as evidenced by the fact that in the Report of the Board of Health of the City of Philadelphia, for 1879, there appear 626 deaths from *convulsions*, for which no cause is assigned ; and remembering how little the physicians of the last and the early part of this century knew about the symptoms of Bright's Disease, does it not seem very likely that some of the cases of death which were then returned as due to sudden or unknown causes, to dropsy or convulsions, were in reality due to this Bright's Disease ? Now, I tell you this in order that you may understand some deductions I am going to make, by which I hope to prove to you, conclusively, that while this disease has probably existed for a long time, yet the number of cases has increased most alarmingly during the last few years.   The earliest record which I can find of deaths in the city of Philadelphia is in the shape of a small, time-worn and stained sheet of paper, about the size òf a page of foolscap, which contains the births, marriages and deaths occurring in *"Christ Church Parish,"* kept by Michael Brown, clerk, and Charles Hughes, sexton, and bearing the date 1740.   Here I find a total of deaths amounting to 165.   There is no mention of any disease of the kidney.   Fourteen deaths are recorded as occurring from convulsions and two from dropsy.   These earlier records are very imperfect and are of very little use, further than to show that this disease must have been very uncommon in those early days, as no mention of it occurs.   Therefore, we will pass by these imperfect ones, and come down to 1807, the first year in which the *"Board of Health"* makes a report.   What do we find?

During the ten years from 1807 to 1817 there is

"no record of any death from Bright's Disease."
But I am going to be liberal and allow that some
of the deaths which were attributed to *symptoms*
which we now know to be prominent effects of
Bright's Disease were in reality due to this disease itself, the existence of which had not been
recognized, for the reasons I have given you.
Thus, during these ten years I find recorded 1651
deaths from *convulsions;* now we will deduct from
this number 1414, which occurred in patients
under five years, because nearly all of them are
sure to have occurred from teething, from disordered digestion, or from some of the ordinary
diseases of infancy and childhood; leaving us 237
deaths from convulsions, the real disease causing
the convulsion being unknown, or, at least, not
given. During the same period I find 463 deaths
from dropsy, cause of dropsy not given; 304
recorded as *sudden,* no cause assigned, and 277
from unknown causes. Now, we will add these
together, and we have a total of 1281 deaths from
causes not satisfactorily stated. Now, both
convulsions and dropsy are oftentimes very prominent symptoms in the last stage of Bright's
Disease, and as I have already told you, death

often occurs *suddenly* in this disease, without any previous intimation of its existence. Therefore, death may have occurred from this disease, when physicians had very little knowledge of its symptoms, and have been ascribed to *Dropsy*, to *Convulsions*, or to *sudden* or *unknown* causes. Let us then allow that ten per cent. of these unsatisfactory cases were in reality Bright's Disease. This is a liberal allowance, and is based on the percentage of deaths from consumption, by far the most fatal of all diseases. Ten per cent. of 1281 will give us $128\frac{1}{10}$ cases, which may be ascribed to Bright's Disease. The total number of recorded deaths for these ten years was 20,316; this would give us a fraction over six deaths in every thousand as due to Bright's disease. Let us now make a big jump and come down to our own times. During the ten years from 1869 to 1879, I find recorded a total of 162,123 deaths in this city. Until the year 1874 there is no mention of *"Bright's Disease;"* but a cause of death is set down in many cases, which is a prominent symptom of this disease and does not occur in any other affection, while in many other cases a different name is given to this disease. For, let me tell you, many of the common diseases have several names, unfortunately, and the physician is at liberty to choose that which suits him. Therefore, these deaths can, beyond doubt, be ascribed to those affections of the kidneys which we know under the comprehensive term of Bright's Disease. For all practical purposes, this name may be applied with good reason to all of them, as they are all capable of being produced by the same causes, are subject to the same hygienic laws, and generally have the same termination. For some reason, unaccountable to me, some physicians are opposed to calling this disease *Bright's* Disease, objecting on general grounds to the naming of any disease after an individual. We name our streets, our cities and our public buildings after distinguished citizens, and I cannot understand why we should not honor the memory of a man who has contributed in a particular manner to the advancement of our noble science, by adding greatly to our knowledge of any particular disease, by naming this disease after him. Of course, for scientific discussion, it becomes necessary to distinguish between the different forms of Bright's Disease, and hence a sub-division of nomenclature becomes imperative; but for ordinary use and for non-professional purposes the name of Bright's Disease seems very appropriate. Thus, some physicians record death as occurring from *Uræmia*, which means a poisoning of the system from the retention in it of the urea which the kidneys should have removed. So uræmia is the immediate cause of death, but the *disease* which has produced the uræmia is *Bright's Disease*, and why not say so.

To go back to my subject: I say I find recorded for the ten years given above a total of 162,123 deaths. Now, of this number 2328 can be set down as due, beyond a doubt, to Bright's Disease, thus giving us a fraction over fourteen deaths in every thousand from this disease. For the benefit of any physician who may read this book, I will give a list of the cases which I attribute to Bright's Disease :—

Albuminuria, ...	161
Congestion of the kidney, . . .	8
Degeneration of the kidney, . .	24
Fatty degeneration of the kidney,	23
Inflammation of the kidney, . .	260
Uræmia, ...	366
Disease of the kidney, ...	572
Bright's Disease, ...	914—2328

Now, if we take the cases of death from dropsy, convulsions and unknown causes, as we did from 1807 to 1817, we will find a still greater increase; thus, during these ten years there are recorded 608 deaths from convulsions in people over five years of age, 1839 from general dropsy, and 588 from causes unknown, giving us a sum total of 3035. Now, if we attribute ten per cent. of these deaths to Bright's Disease, we will have 303 to add to the 2328, making 2631 deaths which, in all probability, were due to Bright's Disease, giving us a fraction over sixteen deaths in every thousand from this disease; thus showing an increase of two and two-thirds as many cases of death from Bright's Disease, in proportion to the total death rate, as we find from 1807 to 1817. As an additional evidence that some of these *unknown* deaths were due to Bright's Disease, let me tell you that the cases recorded as *unknown* decrease as the recorded cases of this disease increase (with the exceptions of 1870 and 1877), as the following table will show:—

	Bright's Disease.	Unknown.		Bright's Disease.	Unknown.
1869	140	90	1874	195	47
1870	*154*	*117*	1875	299	35
1871	182	116	1876	2ợ0	*63*
1872	208	47	1877	2ଃ8	21
1873	211	39	1878	342	13

B*

The apparent exception in 1876 may be explained by the presence in the city of so many strangers, many of whom, no doubt, died very soon after coming under the charge of a physician, before he had a chance to make a diagnosis, and a post-mortem examination being impracticable, the case had to be returned as due to an unknown cause. You will understand this when you read further on and see what my experience was in the Philadelphia Hospital, with this disease. This same relative increase and decrease I find to hold good, with the recorded cases of convulsions in persons over five years of age and of dropsy.

The above table contains another very instructive point. You will notice that the mortality from Bright's Disease in 1869 was 140, while in 1878 it was 342, nearly two and a half times greater, while the total mortality in 1869 was 13,428, and in 1878, 15,743, an increase of only 2315, less than one-sixth.

This is really the most instructive and most incontrovertible point I have given you. There is no theory or speculation about this statement, but stubborn and undeniable facts and figures here show an enormous increase in the proportion of deaths from this disease, even in the last ten years.

I could go on multiplying statistics indefinitely, but I feel sure I have already given you enough to satisfy you beyond any question of doubt that this formidable disease has become much more frequent of late years, and is steadily and rapidly on the increase. I desire to show you this fact and to explain to you the cause of this increase, in order that you may understand the means and realize the necessity of reducing its terrible frequency.

When you have read this little work through, if you think over it for a few minutes, you will realize that all the instructions and advice which it contains can be resolved into the single short precept, *"Live properly;"* live so that every organ in your body will have its own proper amount and kind of work to perform. Now the most powerful cause of this disease is the neglect of this maxim. It is not a so-called inevitable disease, such as some of the fevers, but is much under our control. Exposure, over-work, abuse of our various organs, by over-eating and drinking, and, in a word, neglect of hygienic rules,

will cause it. While, on the other hand, a strictly
proper life in every respect will do much to retard its increase. Suppose it is an hereditary
affection, and may be transmitted from parents to
their offspring; we all know that consumption,
that most particularly hereditary disease, may, in
many cases, be rendered torpid and quiescent,
and its development be stayed, by a proper and
careful life. To point this, let me tell you that
Bright's Disease is much more frequent and more
fatal among men than among women; and I need
not tell you that men are more exposed to mental
worry, over-work, disordered digestion, from ir-
regular eating, excessive drinking, late hours, the
use of tobacco and almost all injurious agents,
than women. Again, it is much oftener observed
among adults than among children, for the same
reasons. I will tell you, further on, that a weak
kidney is like a weak boiler; if you do not put
too much strain on it it will bear the pressure, but
if the strain be too great, it will explode; and I
will tell you that a neglect of the laws of health
will have the same effect on a weak kidney, predisposed to disease, that a steam pressure of two
hundred pounds will have on a boiler capable of

withstanding only a pressure of one hundred pounds, namely, neither one of them can properly perform their duty, and must, beyond question, give out.

No doubt many of the sudden deaths which occurred prior to the time of Bright, and were attributed to apoplexy, heart disease or *unknown* causes, were in reality due to the unsuspected presence of this disease. Because, so insidious is its course in many cases that its existence is not even suspected ; sometimes for many years. Let me illustrate this insidious nature of Bright's Disease by relating to you a few actual cases.

CASE I.—A lady, in apparently good health, who rarely complained, and then only occasionally, of some slight and transient ill-feeling, as nearly every one does at some time or other, took a long walk one day, with her husband, and upon returning home (feeling particularly well), went to her room to remove her bonnet, and while standing in front of the bureau, fell to the floor in a convulsion, followed by twenty-four hours of unconsciousness and death without a return to consciousness. A post-mortem examination revealed Bright's Disease.

CASE II.—A young married man for years suffered from dyspepsia; he consulted many physicians, who prescribed for his indigestion; dissatisfied with the non-success of their treatment, he would consult a second doctor before the first had time to suspect any organic disease as the cause of dyspepsia. After suffering in this way, and having had no other symptoms of disease, for several years, he awoke one morning to find a very slight swelling, scarcely more than a puffiness, under one eye. On his way to the city he consulted a country physician, in the small village in which he lived. The doctor asked him if he had any kidney trouble. He scouted the idea. Well, said the doctor, you had better see your physician when you reach the city, and direct his attention to your kidneys. An examination revealed an advanced case of Bright's Disease, and in less than a month he was dead.

CASE III.—A lady of over sixty, previously in apparently very good health for one of her age, had an attack of rheumatism. For many years she had been afflicted with similar attacks. After the rheumatism left her she seemed exhausted; she did not regain her usual vigor, but had no well defined symptoms of any disease. An examination revealed Bright's Disease, and she died in three days.

CASE IV.—A man of thirty years of age, in apparently good health, complaining of no disease, said, one evening, in conversation with a physician, "Is excessive sleepiness indicative of any disease?" An examination revealed a pronounced case of Bright's Disease.

CASE V.—A married man of thirty-five complained of nothing, yet his wife thought he did not look well; he had a pale, tired look, though he was actively engaged in professional pursuits. She persuaded him to consult a physician, and Bright's Disease was the verdict.

CASE VI.—A young man of eighteen, in apparently fair health, complained of indigestion and general weakness, but no definite symptoms of any form of disease. A visit to his physician resulted in another case of Bright's Disease.

Let me give you one more case. A young man of twenty-two, in previously good health, to all appearances, was taken suddenly sick, and disease of the liver was the diagnosis. He was treated for this for two weeks, and not getting better, I was called to see him. The only symptoms of disease were great weakness, cramps and loss of appetite. His peculiar appearance made me suspect Bright's Disease, and an examination revealed a pronounced case. Careful inquiry from his family failed to elicit any previous symptoms of disease, except a very pale complexion, though the disease must have existed for some years.

During my term as resident physician in the medical wards of the Philadelphia Hospital, it was a common occurrence for an ambulance call to be received from one of the down-town station houses. Upon reaching the place designated, we would be shown a man or woman who had been found unconscious on the street, and supposed to be intoxicated. He would be removed to the Hospital, where he would linger unconscious for twenty-four or thirty-six hours and then die. In the majority of these cases a post-mortem examination would reveal Bright's Disease.

I have given you these instances because I wish to show you that very often this disease may exist and yet the symptoms which it presents to the person so afflicted may be so slight as to cause him to neglect professional advice.

Let me here, in the beginning of this little book, enunciate a very valuable rule. If you ever experience a departure from *perfect* health, no matter how slight this departure may be, and if your symptoms do not indicate disease of any organ, but seem merely to be the temporary result of some excess in eating or drinking, work or exercise, or some slight exposure ; if these symptoms continue for any time after the removal of the cause, ask your physician to make an examination of the condition of your kidneys. If they are not diseased, so much the better ; if they are, it is well for you to know it at once, because I will tell you now (and tell you why later), that a person with weak or diseased kidneys can, by leading a careful life, enjoy very fair health, perform a reasonable amount of work, and attain a respectable degree of longevity ; while on the other hand (as I told you in my preface), very few diseases are so strongly influenced for the worse by a reckless and careless life as the one under consideration.

PART SECOND.

THE FUNCTIONS OF THE KIDNEYS AND THEIR DERANGEMENTS.

In order that you may understand and be induced to follow the mode of life here recommended, I will now tell you, in plain language, divested of all technicalities, some little about the kidneys and the duties which they *ought* to perform.

The kidneys, two in number, are situated in the *small* of the back, on either side of the backbone. They are small, but most exceedingly important organs. If they were removed from the body it would be utterly impossible to support life. They are among the principal scavengers of the human body. You all know that particles of our bodies are continually dying, their places being supplied by new particles, resembling the old ones, and derived from the nourishment which we take. Now, these dead particles undergo a process of decomposition in the blood, just as our bodies in mass will do after the final death, and are then removed from the blood through the agency of certain organs, which act as purifiers, so to speak, removing all dead and decomposed particles from, and thus purifying, the blood. Prominent among the products of this death and decay of tissue stands a substance called *urea*. The chief duty of the kidney is to eliminate this *urea* from the blood and carry it out of the body as the principal ingredient of urine. This is the only function of the kidney with which we have to do, as all the trouble in Bright's disease may be principally attributed to the non-elimination of this *urea* and to mechanical interference with the circulation of blood in the kidneys. Now, if disease so interferes with and cripples these organs as to prevent them from removing the *urea*, what becomes of it. The death and decay of tissue continue, whether the kidneys be weak or strong; *it* must go on as long as life, for it is this very destruction and renewal that constitutes life. When the renewal ceases and the destruction predominates, decay of life and death result. So, with the continued destruction of tissue, there is continual formation of *urea*, and if it cannot be removed from it must accumulate in the blood. Now, everybody knows that the blood circulates throughout the entire body, and if contaminated with *urea*, it must carry this poison to every tissue which it nourishes, and so impress them unfavorably by its presence. That is just exactly what it does do. Then, no doubt, you will wonder why such general poisoning does not produce more marked symptoms, and why the existence of a disease which causes such general evil is not made more manifest. This is due to the law of tolerance. A perfectly temperate man will be affected by one drink of liquor ; by degrees, he can drink more and more without feeling it, until eventually he can consume enormous quantities without apparent effect. So it is with everything else. A continued use of any poisonous agent, commenced in small doses and gradually and slowly increased day by day, will eventually breed such a tolerance that enormous quantities, a fraction of which would prove fatal to the novice, can be taken with impunity. I once had a patient under my charge (who, by the way, was suffering with Bright's Disease) who could and had taken daily sixty grains of morphia, with but little effect, when

I should hesitate to order half a grain for one unaccustomed to its use. So with the accumulation of the urea. When the disease in the kidney commences the integrity of the organ is so little affected that it performs its duty *almost* perfectly, and only a very small quantity of urea remains in the blood. Day by day, as the disease advances, the quantity of this poison that is retained slowly increases, the system becoming, in the meantime, accustomed to its presence, and presenting no marked symptoms of revolt against its unwelcome tenant. Finally, the point of forbearance is reached and passed. The large accumulation of urea in the blood explodes, as it were, to use a familiar word, and we have convulsions and death. You can now understand why I call Bright's Disease an insidious affection. You can also appreciate how apt a person would be not to consult a physician, imagining his bad feelings to be merely the result of fatigue. You can, at the same time, see how a course of life which would throw as little work as possible on the kidneys would tend to longevity. I once heard an eminent physician compare the situation of a man whose kidneys are diseased to a ship at sea which has sprung a-leak.

As long as she encounters fair winds and weather she is all right, and will probably reach port in safety, if her pumps hold out. But let her meet a severe storm, or let her crew become too much exhausted to man the pumps, and she will surely founder. So it is with a man whose kidneys are diseased. So long as he leads a careful life putting but little strain upon these weakened organs, so long may he live in very fair health and comfort; but let him be imprudent and reckless, let him pour a tempest of abuse and neglect upon his poor kidneys, and they will surely succumb·to the strain.

PART THIRD.

ABOUT BRIGHT'S DISEASE (WHAT IT IS).

Bright's Disease, in all its forms, is essentially an inflammation of the kidneys. They contain more blood than in health. In some forms the kidney, at first enlarged, afterward becomes contracted. Its tissue, pressing on the numerous blood vessels which pass through the organ, interferes with the free passage of the blood through them. This backward pressure on the current of blood being met and opposed by the onward pressure of the powerfully acting heart, the watery constituents of the blood are forced through the porous walls of the blood vessels, usually of the feet and legs, and so we have the dropsy produced which is frequently present in the advanced stage of this form of the disease. Again, this backward pressure causes so much strain on the heart in its efforts to overcome it, that we sometimes have enlargement of the heart as a secondary effect of Bright's Disease. This pressure will sometimes cause the watery parts of the blood to ooze out into the small cavities in the lungs, usually filled with air, and death will ensue, from a gradual suffocation. Let me here digress, to tell you that this mode of death is not attended with great suffering, as many would suppose ; on the contrary, it is painless and even pleasant. I will tell you why. In the first place, let me tell you that all pain is felt in the brain. If you cut your finger, the nerves of sense, immediately and with the rapidity of lightning, convey the news to the brain, and the pain is there experienced, though the sensation of pain is referred to the seat of the injury. To make this still clearer, if you had no brain your body might be cut in two and you would not feel the slightest pain, even though you had life, for it has been demonstrated by experiment that life is possible even after the removal of certain portions of the brain in which sensation resides. Secondly, in order that the brain may be able to receive sensations of pain, or of any kind, it must be in a state of activity, and to be so it must receive good nourishment, from pure blood, containing a sufficient proportion of the life-giving oxygen. Now, when this water oozes out into the cavities in the lungs in which the air usually is, you can understand that the oxygen cannot get into the lungs, and so cannot enter the blood. Now, as the particles of our body die and decay, some of them assume the form of carbon, and as such are carried in the blood to the lungs, where they meet and unite with the oxygen to form carbonic acid, in which shape they are carried out of the body in expiration. If oxygen cannot get in, of course, carbon cannot get out, and must remain in the blood, to poison it. Nearly every one is familiar with the effects produced by inhaling the fumes of charcoal. Well, here we have the same thing. At first, when the water commences to ooze into the lungs, a very small amount of carbon only is compelled to remain in the blood, and being carried to the brain by the blood, it serves to stupefy it, as it were, to render it incapable of feeling pain, only, however, to a slight degree. As more water oozes out and less room is left for the oxygen, of course more carbon remains in the blood, and the brain is rendered more torpid, less sensible of suffering, until finally, when the difficulty in breathing has reached that point which would

c

prove very painful to a person in health, the brain is so overcome and clouded by the carbon that the patient fails to feel any suffering, though to those standing around he may seem to be suffering most intensely. The urea retained in the blood may accumulate in such enormous quantity as to finally poison the brain and cause convulsions, or acting more gradually on the brain, may produce unconsciousness, blunting one sense after another, as it travels from the upper portion of this organ downwards, until it finally reaches and contaminates that portion of the brain from which arise the nerves that convey the power to the heart to move, and to the lungs to breathe, and the unconscious and insensible patient quietly ceases to breathe, his heart stops pulsating and the curtain noiselessly drops on his drama of life.

I have now hastily given you the functions of the kidneys; how they are affected in Bright's Disease; the result of these irregularities in action, and the usual modes of death. I now come to the kernel of this little work. I will now tell you how a person threatened with this disease may postpone its development, and how one in whom it has already appeared may retard its progress. First, let me again repeat what I have already said : that this disease is greatly influenced for better or for worse by our own actions. We have, to a certain extent, in our own hands the power to shorten or prolong our lives. The person with Bright's Disease may, by pursuing a correct life, outlive thousands of those around him at present in vigorous health. He may live many years and ultimately die of some other disease. Neither is it necessary that he should live the life of a confirmed invalid. He can work and can enjoy life, within certain bounds, as well as any one else. But he must always be careful. The object of his life must be to put as little strain as possible upon his kidneys. Further than what I have already said of the symptoms of this disease I will not go. The diagnosis comes within the province of the physician, and it is not my intention to supplement him. A good physician is a great gift bestowed upon man by an all wise and benevolent God ; and let me here strongly advise you to immediately consult some *good* doctor upon the first intimation of the approach of this disease, and to follow his advice implicitly. So, if any of my readers have felt unwell, low-spirited and weak, for some time, without any definite symptoms, let them find out whether or not they have Bright's Disease.

This disease may be divided into the acute or rapid, and the chronic or slow forms. The acute type we have nothing to do with here, because its symptoms are usually so marked and well developed as to require the calling of the physician. It rapidly runs its course and terminates either in death, complete recovery, or by degenerating into the chronic form, the special subject of this little book.

PART FOURTH.

RULES OF LIFE.

I hear some people asking me what I mean by a *predisposition* to Bright's Disease. It has not yet been set down in medical works as an hereditary affection; though I believe the day is not far distant when it will be so considered, just as consumption and scrofula are at present. I will illustrate my meaning by an example. I know a family in which both parents and two sons have died of this disease; of four surviving sons, three, and of two daughters, one, are now afflicted with it. Two young children of one of the sons who died have it. Now, I would say, that in all human probability those members of this family who have not the disease already, have a predisposition to it; their kidneys are their weak points, so to speak, and any deleterious causes acting on their systems would probably produce kidney disease. Hence, these members should so live as to protect their kidneys as much as it lies ·in their power.

They should avoid a sedentary life as far as possible, as well as mental worry, and should indulge in mental work or brain effort only to a moderate degree, preferring an out-door, active life, and plenty of exercise and motion in the pure, fresh air, to an occupation confining them to the foul atmosphere of a city house. They should, however, carefully avoid overtaxing their physical strength, for while excessive mental action is very injurious, excessive physical labor is only a little less so. Many intelligent people unconsciously fall into the very natural error that great physical exercise is beneficial to health, and that the more they take of it the better. While this is true, within bounds, as far as a person in vigorous health is concerned, it is not so for one whose kidneys are diseased, for the following reasons: It is one of the functions of the kidneys to remove the results of the death and decay of that certain class of tissue whose predominant constituent is nitrogen. Now, muscular tissue is very rich in this element. Every motion of a muscle causes the death and decay of some of its component parts; hence, you can understand how excessive muscular action will result in an excessive production of dead and decayed elements, rich in nitrogen, which must be removed by the kidneys. So you will see that the work to be performed by these organs will be in proportion to the use which we make of our muscles. At the same time, a moderate amount of muscular exercise will give tone to and improve the general system, will exhilarate the circulation and keep the skin in proper action, the importance of which you will see further on, and will thus redound to the benefit of the weak kidneys. Hence, I would enunciate, as a general rule, applicable to the great majority of cases of Bright's Disease, where the strength has not been too much reduced, that while a walk of three or four miles will be of positive benefit to the kidneys, one of ten miles will be injurious, as it will throw too much labor on them in removing from the body the decayed products of this amount of muscular exertion. Excessive mental work is forbidden in this disease, not so much because it is injurious in itself to the weak kidney, as on account of the sedentary, in-door life, its usual accompaniment, which, by depressing the general system, reacts unfavorably on the kidneys. A considerable amount of brain work, performed in the open air; varied with a proper amount of physical exercise, can be borne with impunity. Persons afflicted with this disease should endeavor to select a healthy medium in both. Let me suggest a daily routine, which will probably come as near being correct as possible for the average man or woman fond of mental work. Suppose he should arise at seven o'clock : let him breakfast and leisurely read his morning paper. Then let him devote himself, say from nine until twelve o'clock, to his mental work ; at the end of each hour intermitting his labor for five minutes' light exercise. Then, for two hours, until dinner time, let him amuse himself with some *out-of-door* occupation. After dinner and a short period passed in pleasant conversation or *light* reading, so as to allow digestion to fairly commence in peace and quietude, let him drive, ride, or walk, or engage in some *out-door congenial* exercise, from half-past three until five or half-past ; then to his mental work for an hour, until supper, at half-past six. The evening to be passed in conversation or light reading. Go to bed as near ten o'clock as possible. Be sure to secure eight hours of sleep out of the twenty-four. To the earnest and ambitious brain worker four hours' labor daily

will seem very inadequate to accomplish what he
desires; but to such a person let me say, for his
comfort and peace of mind, that, leaving all considerations of preservation of health out of the
question, which really is the point we are discuss-
ing, the man who works *earnestly* four hours a
day, with his mind vigorous and free from mental
cobwebs, will accomplish *more* satisfactory work
in the aggregate, than he who labors twelve or
fifteen, with his mind weak and exhausted and
his ideas confused, from over strain. The rest and
exercise will refresh and rejuvenate his mind, and
he will, each day, go to his work with a relish for
it; his perception will be keen, his mental appetite will be good and the *assimilative powers of
his brain* will be vigorous and capable of receiving
and storing away in its proper place, without con-
fusion of ideas, all that his mental stomach may
take in. For one reason or another, it would be
next to impossible to induce a man to live by rule,
and particularly by the rule of another, but for
the sake of that health which we all *prize so
highly* (when we do not possess it), let me beg my
fellow creatures to catch my idea of leading a
regular, moderate, temperate life in everything.
    c*

The whole secret of longevity (whether you are sick or well) lies in this moderation, temperance and regularity; and it is to show you in a plain way how to so live that I have undertaken this little work.   Let me point what I have said by an illustration.   Case No. 5, already quoted, occurred in a professional man.   As soon as the presence of the disease was discovered his physician insisted that he should abandon his work and devote himself to the care of his health.   Although an earnest lover of work and very ambitious, he followed this advice.   He passed one winter in Florida, leading a life of recreation and amusement.   He would drink sulphur water and bathe two or three times a week in the sulphur pool. The only medicine ordered was a general tonic for the whole system, to build up and improve the tone of the blood.   He improved very rapidly. In a short time his appearance improved so much that his companions would ridicule the idea of his being sick, and called him the healthiest looking invalid they had ever seen.   He slept well, ate well, digested well and enjoyed life as much as the most vigorous man.   He returned home in very good condition.   For some three years and a half he continued to follow his physician's advice. He worked but little, took much recreation, spent a great deal of his time in the open air, and continued to enjoy very good health. Finally, being induced by his continued well feeling to believe that his kidneys were in reality much better than they were, and being tempted by the prospects of great profit, he embarked on a business project which entailed great work and much mental anxiety and worry. In less than four months' time he fell back to where he had been three years before; in a short time his feet and legs began to swell, and after a few months' confinement to the house he died. Now, in this case his three years and a half of careful life did not really improve the condition of his kidneys; it did not remove any of the degeneration which had taken place, but it reduced to the minimum the amount of work which these organs had to perform, and so enabled him to live a comfortable life. Neither did he live the life of a confirmed invalid; he enjoyed himself, and was careful merely to avoid excesses of all kinds, exposure and over work. Had he continued this mode of life he .might have lived many years. You see how rapidly an opposite course of life produced disastrous results. Let me here enunciate one great rule: "It is not mental *work*, but mental *worry* that kills." In the case just quoted the business venture referred to caused much mental anxiety, while the patient himself was of that nervous temperament, so that small matters worried him exceedingly. I could point you to a case where the patient has had infinitely more cause for mental worry than the one referred to, but being of a sanguine temperament the effect was not so marked, and he is to-day living, in fair health, though he has had kidney disease for some years. Still, however, the rule holds good for the sanguine as well as for the nervous temperament, that mental worry is particularly prejudicial and hurtful to the weak kidney. In the different temperaments, of course, the effect will vary in the rapidity with which it becomes manifest, but does not vary in the fact of its production. Most naturally, many persons will say: But if we have nervous temperaments and have cause to worry, be it of a business or a social nature, we cannot help but worry; we cannot control our feelings. Here you make a grave mistake. Your tempera-

ment and your feelings are more under your
control than most of you suppose. It is true
that your physician cannot give you any *drug* to
control your temperament, but *moral* medicine, so
to speak, here comes into play, and is a most
powerful agent in enabling us to mould our dis-
positions and control our natures. Thus, without
being a fatalist, it is very easy and very comforting
for every one to believe that whatever may happen
to them happens for the best. It is only necessary for them to believe in the existence of an all
wise and all powerful Creator of everything and
everybody; one who has the wisdom and the
power to ordain that everything which happens is
for best. Such a belief will do much towards pro-
ducing a tranquil frame of mind, so important
and so necessary to the maintenance of health
and the promotion of longevity, and doubly, aye
twenty times, more important in the disease under
consideration. I cannot better describe to you
what I mean by my *moral medicine* than by
quoting to you from a letter which I have
lately received from our distinguished and vener-
able fellow citizen, the Hon. Eli K. Price. It
came to me in answer to a letter addressed to

him, asking for some of his rules of life, by
which he had been enabled to reach such a degree
of longevity and of activity, which I desired to
use for another work I have now in course of
preparation. He says, "A tranquil mind I find
a requisite to health and longevity; and that the
mind may be stayed in a firm tranquillity, I have
found a firm faith in an overruling Providence
essential. My religious belief, insuctively formed,
has all the conviction of my understanding, as
strong as that in all established scientific truths.
All truths have come from the Creator, and all are
his. God's attributes of might, justice and love,
we must logically believe; and without these and
His absolute goodness we must infer Him and His
creation to be failures. It is in believing Him to
have these attributes; in believing that He appeared in Christ to save immortal souls, and for
that visits them, we can endure, and only so
endure, the sufferings incident to life with firmness
and equanimity, often emerging from afflictions
with safety where others fail; and thus fortified
we may humbly hope and expect to reach the end
in a faith that will be sustaining, and to pass into
happiness, not less, but the greater for that well

doing which had made this life happy." Further
on he says, "Some hours of every week, for
twenty years, have, with pen in hand, been devoted
to the study of the being of God, his attributes,
and the significance of his works; of the nature
and import of our being and destiny; until the
product has become voluminous. Hence, I speak
with the confidence of the absolute conviction
that has sustained me through nearly all of life,
that man has the help of a care and strength
greater than his own. With such faith and 'a
love that casteth out fear,' trials melt away before
us; we live through them and see them as clouds
that have passed, dispensing blessings, and in a
hopeful confidence in God's care and love we live
longer, better and happier, and even in death
know a triumphant joy." I have given you the
convictions of a gentleman, who being in youth,
as he tells me, not particularly robust, yet has
passed through a long and honorable life of great
mental labor, and is, in his eighty-fourth year, still
hale and active, engaged in the ordinary pursuits
of life, taking an interest in all that occurs in the
world around him, continuing to give legal opin-
ions, and, in a word, is more useful than most men

of sixty.　Now, after reflecting on what he tells us, can any reasonable man still insist that a tranquil frame of mind is impossible when he has cause to worry.　Let every one try, to the utmost of their ability, to do right, so far as the knowledge of the right is in them, and then, if things go wrong, let them attribute this *apparent* wrong to the all-pervading wisdom of the great Creator, and believe it to be for the best, and surely a tranquil mind and freedom from mental worry will be the result.

Any one suffering from, or predisposed to Bright's Disease, should reside in the country, where the air is pure and uncontaminated with the foulness of a large city.　At the same time they should carefully avoid draughts and exposure to wet, damp and inclement weather, to which dangers people in the country are particularly liable. In choosing a residence in the country it should be borne in mind that this disease occurs most frequently in humid, marshy climates and on the sea coast, and that variableness of climate favors its production.　Hence, we should select an even climate, in a dry, elevated and inland locality. Very cold weather should be avoided, because in a person whose kidneys are diseased the skin takes the place of the pumps in the leaky ship. Anything which interferes with the free action of the skin will impress the kidneys unfavorably. A fact often noticed will demonstrate the truth of this statement. Among the poor it has been observed that want of cleanliness of the skin is a powerful predisposing cause of this disease. I will tell you why. A large amount of water is daily thrown out through the pores of the skin, and in this water there is a certain amount of urea. So you will be prepared to understand when I tell you that there exists a very friendly relation between the skin and the kidneys. When the skin throws out a large amount of perspiration, as in warm weather, there is a correspondingly smaller amount of work for the kidneys to do. On the other hand, in winter, when the pores of the skin are contracted by cold, and less perspiration is given out through them, the kidneys have a larger amount of work to do. Let me give you a marked illustration of this fact. In Case No. 2, quoted in the early part of this book, the disease was discovered in November. The patient was immediately placed under treatment. For some time he remained about stationary, apparently no worse and no better. On a certain evening, complaining very much of headache, dry cups were applied to his neck, to draw the surplus of blood from his head, and means were employed to promote the action of his skin. The effect was marvelous ; he slept soundly all night (which he had not done for weeks before), and awoke in the morning feeling, as he expressed it, splendidly. He ate a very hearty breakfast and digested it, and said he felt perfectly well. It was a cold December day. Having some business, he ventured out. He had not been in the street half an hour before he became so sick that he was compelled to return home. He grew rapidly worse, had a convulsion in less than forty-eight hours, and was dead in less than a week. Now, you see, the free action of the skin which was brought about removed from his body the poisonous urea and allowed him to feel well, and his stomach to digest a hearty breakfast. As soon as he was exposed to the cold air the action of his skin was checked, his kidneys were too much diseased to remove the urea, hence it accumulated in and fatally poisoned his system. One of the most perfect illustrations of this correlative action of the skin and kidneys that can be imagined .has just occurred to me while writing this little book. I was examining an obscure case, supposed to be liver disease, and which I ultimately discovered to be Bright's Disease. In making this examination it was necessary for me to handle the patient a great deal. Upon its completion I was unable to thoroughly wash my hands. Some two hours, or more, afterwards, I accidentally put my hand near my nose, and was immediately struck by the strong, pungent odor of urea, clearly showing the surface of the patient's body to be coated with it. In other cases I have noticed this odor given off from the skin in a person very far advanced with Bright's Disease. Now, please understand me. I do not recommend you to keep yourself constantly in a violent state of perspiration. If you do, you will do yourself a great deal of harm. What you want is to keep up that insensible perspiration, as it is called, which in a healthy person is constantly taking place. Let me tell you that about a pint of water is given out by the skin every twenty-four hours, even when it does not feel moist to the touch. This water is absorbed by the surrounding atmosphere as soon as it appears on the surface of the skin, so that to ordinary observation its presence is imperceptible, hence it is called *insensible perspiration.*   Now, in order to keep the skin in proper action, we must attend to several points.   We must particularly avoid draughts and cold weather ; we must always be warmly clad.   You all know that heat expands and cold contracts ; hence you can understand that cold weather, by its contracting influence on the skin, will prevent the free passage of water through its pores.   The clothing worn next the skin should be of a porous texture, because it will absorb the water as it is given out.   If the clothing does not take up the water, but allows it to accumulate on the skin, it interferes with its removal, and consequently with its elimination by the skin. So, I tell you, wear *woolen* and do *not* wear *cotton* under clothing.   Let me point this precept by relating a case which has just occurred in my own experience.   While writing this very chapter on the avoidance of cold and the clothing to be worn in this disease, I was called in haste to see a patient who was reported to be very sick.   Upon reaching his bedside I was immediately struck with the peculiar pale, earthy appearance of his face, which is so characteristic of this disease in its advanced stages that it might with propriety be termed the *"Brighty Complexion."* He had for some time been treated for disease of the liver. I made an examination of his urine and discovered a marked case of Bright's Disease. When I made known the nature of his disease to his family, I was questioned by his father as to the cause. I told him the causes were numerous ; among others, I said, is exposure to cold and an insufficient amount of clothing. "Ah!" said the father, interrupting me, " that is it. Why, on the coldest day in winter I could not get this boy to wear drawers ; he thought he was so strong and well that he would expose himself in the most inclement weather, improperly clad ; and would never think of changing his clothing after getting wet. Working in a mill, he would often, when in a violent perspiration, sit down anywhere, not thinking or caring whether he was in a draught or not." One of the most marked cases of this disease that I have ever seen occurred in a man who was employed on a steamer as a stoker, a class of men among whom this disease is very prevalent. I need not tell any one who has ever been on an ocean steamer that these *stokers* will spend several hours in the hold of the vessel, stripped to the waist, shoveling coal. They perspire most violently, and when their term of work is over, still stripped, they hurry to the deck to cool off. Any one can foretell what occurs. The perspiration is suddenly checked by the strong wind and cold or cool weather always found at sea; their expanded pores are contracted, hence, extra work is thrown on their kidneys. At the same time, the blood in the vessels of the skin is driven out of them by the contracting influence of the cold, and forced into the internal organs in undue amount ; and if a latent tendency to inflammation of the kidneys should exist, it is brought into activity by these means.

It is of the greatest importance to bathe frequently in this disease, because the upper layer of the skin is constantly dying, and its dead scales adhere to the body, and if allowed to remain they will offer a mechanical obstacle to perspiration. Again, the *insensible* perspiration of which I have told you carries with it particles of dead and decayed tissue; the atmosphere absorbs the water, while the decayed particles are left on the skin. If allowed to accumulate there they will form a coating, which will interfere with the passage of the water. I would recommend you to take a daily bath, and probably the best time would be just before getting into bed. Because, if you choose the morning you will be apt, sometimes, to go out into the cold air with the pores of your skin wide open, from the bath, and the air suddenly contracting them will throw a strain on the kidneys. *Warm* baths are to be recommended, because they tend to open the pores of the skin and promote its action. A warm bath is also a sedative, and tends to produce sleep, which, by the general rest it gives to all the organs of the body, will be favorable to the weak and exhausted kidneys. Cold baths are to be condemned, because, by contracting the vessels in the skin, they force the blood out of them, and cause it to seek a home in some of the warm interior parts ; the kidneys being diseased are, of course, weak, and offer the least resistance to its unwelcome intrusion, and so it lodges in them, congesting them and adding fuel to the flame. If you bathe daily, do not remain in the bath more than ten minutes ; if you do you will exhaust and weaken yourself. If you find a full bath too

weakening when indulged in daily, make use of a
bath by means of soap, water, and a towel, only
exposing a portion of the body at a time, and
covering it up as soon as washed, or you may take
cold. A sponge bath I can only characterize as a
filthy abomination. A sponge will absorb and
retain all manner of dirt, and it is next to impossible to keep it clean. To believe this, smell a
sponge after it has been some time in use.

Turkish baths, occasionally indulged in (about
once a week), will do much toward promoting
healthy action of the skin, but let me emphasize
one point. In cold weather remain in the cooling-
off room and do not expose yourself to the outside
air for a good hour, until you are thoroughly
cooled. By neglect of this precept the bath will
do more harm than good, as the cold acting suddenly on the expanded pores will so thoroughly
contract them as to throw more work than ever
on the weak kidneys. For the same reason I
would omit the cold plunge at the end of the bath
and substitute a lukewarm shower bath. Now,
do not misunderstand me and imagine that your
skin can be made to take the place of your kidneys. Not by any means. A man whose kidneys

are worthless and unable to remove urea cannot live, no matter how healthy may be the action of his skin. There are many people in whom the kidney disease has only advanced to the stage of inflammation, where there is no degeneration of the structure of the kidney. Now, it is an all-important axiom in the treatment of inflammation, wherever it may be located, that the inflamed part should be relieved as much as possible from work, and allowed to rest. So, in these cases, you can understand that if the skin is kept in healthy action it will help the kidneys to do some of their work, and will thus give them a certain amount of rest and enable them to regain strength and tone. Again, even when degeneration of the kidney has set in, it may be only to such an extent that these organs are able to remove a considerable amount of urea, but not all; here the healthy skin action will remove the balance, which the kidneys are unable to do. Thus, by purifying the blood and so furnishing good nourishment to these organs, their degeneration will be retarded. I need hardly say that healthy skin action is very dependent upon frequent change of clothing. If the clothing be very dirty some of its impurities

D

will adhere to the skin and we have a mechanical
impediment to perspiration. Muscular exercise
tends greatly to keep up healthy action of the
skin, but when heated from this exercise be very
careful to avoid the slightest draught, or by suddenly contracting your skin you will do yourself
harm. Let me here (at the risk of tediousness
from repeating) impress upon the person with
Bright's Disease that *the* most important factor
in prolonging life and enjoying comfortable
health is *healthy skin action.* Because the skin,
more than any other organ or part of the body,
has the property which enables it to help the
kidneys in the performance of their duty. Remembering what I have said, you will understand
why I would advise any one living in a region
where the winters are severe to leave home upon
the first approach of cold weather and spend the
winter months in a warm latitude. Let me impress
this on your memory by the aid of statistics. In
the five years from 1875 to 1880 the deaths from
this disease recorded in the city of Philadelphia,
during the months of January, February and
March amounted to 429, while in the same years,
for the months of June, July and August, I find

only 349. In this country, Florida has for many years ranked high as a winter resort for invalids, and deservedly so. The climate there is delightfully pleasant and the temperature is very even ; the sudden changes from heat to cold and the reverse, so common in our Northern States, and so particularly injurious to a weak kidney, are unknown there. There can also be found in that State sulphur springs for drinking and pools for bathing, which tend greatly to promote healthy action of the skin.

It is obvious, from the facts already given you, that any one having weak or diseased kidneys should exercise the greatest care in reference to their clothing. They should always be warmly clad, and above all things, should immediately change all of their clothing after getting wet from the rain or otherwise, being particular to rub the whole body with a coarse towel, until it glows, before putting on the dry clothing. A simple, and I believe very good, practice for people with weak kidneys is to make use of counter irritation over the region of these organs, morning and evening. This may be easily done by taking hold of the ends of a coarse towel with each hand, and drawing it rapidly

several times across the small of the back, until a
sensation of warmth or slight burning is felt;
persons with tender skin should be careful not to
use a very coarse towel, or they will abrade the
skin and make themselves sore. Let me here explain to you how counter-irritation acts, so that
you may understand the *modus operandi*, whenever I may have occasion to recommend it.
Bright's Disease in its various forms is essentially
an inflammatory affection. Now, the chief and
invariable feature of inflammation is an excess of
blood in the inflamed part. When we put a mustard plaster (a form of counter-irritation) on the
skin over the chest for a pain in the side, what
effect do we produce? By this counter-irritation
we cause the blood to be drawn from the nearest
organ or part to the skin where we have the mustard, and by this means we remove the excess of
blood from the inflamed part to the skin, where it
is harmless, and so relieve the affected organ. So
by this friction with the coarse towel over the
kidneys, we produce the same effect as with the
mustard plaster on the chest; we draw the excess
of blood from the congested and engorged vessels
of the kidneys to the surface, and by repeating

this irritation morning and evening, we help to overcome the tendency to engorgement, and assist the free and healthy circulation of the blood through the kidneys. For the same reason, you should be particular to keep your feet warm. Cold feet indicate an insufficient amount of blood in the vessels of the feet. Now the whole body contains just as much blood, ordinarily, when the feet are cold as when they are warm; where then is the blood that ought to be in the feet? Being driven out of the vessels there by the action of cold, it has naturally sought the weakest organ, where the least resistance was offered to its entrance, and has accumulated in undue quantity in the kidneys, where it wreaks its revenge upon the man who did not have sense enough to keep his feet warm, and drove the blood whose proper home was in his feet out into the world among strangers who did not want it and whose house was already comfortably full. I once knew an eminent clergyman, a man of great learning and sound common sense, who kept constantly on hand a number of stockings of different degrees of thickness, and who, upon rising in the morning, consulted his thermometer (hanging outside of the window) as to which pair of stockings he should wear. I remember, at the time, thoughtless people laughed at him for what they termed his "*old-maidish habits;*" by so doing, they only exposed their own ignorance, for this good gentleman, who attained a ripe old age of great mental work and usefulness, clearly showed that he understood and thoroughly appreciated one great secret of health, *keep your feet warm.* It would be well for people with weak or diseased kidneys to wear over the region of these organs a fold of flannel (two or three thicknesses) about six inches wide, during cold and damp weather. Tape sewed to either end and tied in front over the abdomen, will keep it in position. This will help to protect the kidneys from the evil effects of cold. But let me caution you that, having once commenced its use, never go without it while the cold spell lasts, as the very fact of this extra protection would render the part more susceptible to the bad action of the cold, if its use were intermitted.

Alcoholic liquors should be absolutely discarded in this disease. Their use cannot be beneficial, and in the large majority of cases will be absolutely injurious. Gin is popularly supposed to

have an action on the kidneys, and so it has; it
promotes their action and increases the quantity
of urine; but its use is only safe to tide over great
emergencies. It is like giving the spur to a tired
horse; he will go faster for a time, but the subse-
quent exhaustion will be all the greater. Gin,
therefore, is a diuretic, a medicine that will increase the action of the kidneys. Remember
what I have said about Bright's Disease being of
an inflammatory nature and read what Dr. Black,
of Glasgow, says about the use of diuretics:
"Diuretics are actually employed at the present
day in the treatment of acute and chronic nephritis (Bright's Disease), on what basis in reason or
common sense I have ever failed to comprehend.
I have always regarded it as an axiom, in the
treatment of inflammation, that rest to the inflamed organ is of the first importance. As well,
in my opinion, command a man suffering from
double pneumonia to leave his bed and take a
three-mile race, as to give stimulating diuretics in
a case of nephritis." Now, why is rest so important? I have told you that an excess of blood
in the part affected is the chief and invariable
feature of inflammation. Any motion or exercise

of a part will cause a destruction of some particles of tissue which have been concerned in the motion, and new particles are deposited by the blood in the place of the old ones. Now, if excessive motion òr exercise takes place, you can understand that great destruction of tissue must ensue, and as a natural sequence, an excessive amount of blood must rush into the organ or part, in order that a sufficiency of nourishment may be furnished to properly supply the waste and keep up the balance between formation and destruction, the *sine qua non* of healthy life. So if an excessive amount of blood rushes into any part or organ, we have *the* important element of inflammation; if this influx of too much blood is frequently repeated, the vessels finally become distended and too much blood is all the time in the organ, giving us the condition of *chronic inflammation*. This excessive destruction of tissue is not confined to the voluntary muscles or those under control of your will, as many suppose, but can and does occur in any organ or part of your body. Too much thinking, or seeing, or smelling, or hearing, will cause an excessive destruction of brain tissue. To illustrate this, you are all familiar with the fact that excessive application of the mind, as in study, will cause headache, which is simply due to the fact that excessive exhaustion of brain tissue has caused an extra amount of blood to flow into the brain to repair this waste. The vessels being dilated and rendered larger by this surplus of blood, exert an unaccustomed pressure on the surrounding sensitive brain tissue, and a headache is the result. So if you use any agent which has the property of stimulating the kidneys to increased or excessive action, you are likely to do harm. As you see, you will direct towards the kidneys an unusual amount of blood in order to repair the destruction which the excess of work has caused, but when the necessity for this extra supply has passed away, the vessels of the kidneys being weak, are unable to drive the excess of blood out of them and it remains in a state of chronic inflammation, engorgement or congestion, and we ultimately have degeneration of the kidneys and all the resultant phenomena of Bright's Disease. Alcohol is stimulating to the kidneys; hence you can clearly understand how terribly dangerous its use would be if your kidneys are weak. In many cases of Bright's Disease, there

D*

is so little suffering until towards the last, the
patient being able to move about and join in the
ordinary movements, pursuits and pleasures of
life, as one in vigorous health, that he fails to
consider himself an invalid at all, and is very
prone to be guided entirely by his inclinations in
his mode of life, rather than by his physician's
advice. If sociably inclined, he is very apt to
daily consume a certain amount of alcohol, imagining that it does him no more harm than it
does his vigorous neighbor. Because so many
fall into this fatal error, I intend to go into the
question of the use of alcohol in this disease some-
what at length, and to show you how very insidious and devil like are the evil effects of alcohol,
and how little you can judge of the extent to
which these effects have gone, by your feelings.
First, let me tell you some of the effects which
alcohol produces on the system at large. You all
know that it is irritating ; whoever among you has
ever swallowed a glass of whiskey, has felt the
biting, the burning in the throat and stomach,
which it produces. This burning and sensation of
warmth is due to the fact of the alcohol irritating
the delicate coating of the stomach and causing

more blood to flow into it, just the same as mustard will do if applied to the skin. If frequently repeated, this excess of blood becomes permanent, and a chronic inflammation of the stomach is the result. This condition may be produced in a person who has never been intoxicated in his life, but who has used alcoholic liquors in *moderation* for some time. You will believe me, when I tell you, that a stomach in a state of chronic inflammation cannot possibly digest food thoroughly, and so half prepared nourishment is furnished to the weak kidneys, when they ought to have the purest and best, just as a weak man requires to be fed upon what is most easily assimilated. Again, it is a well established physiological fact and one which common sense will make apparent to everybody, that when vessels are filled to their utmost capacity they are incapable of holding any more; hence these vessels in the walls of the stomach, which, properly acting, should take up some of the products of digestion, are so engorged with blood as to be unable to do so, and as a natural consequence, a portion of that which was intended to repair the wear and tear of our bodies is unable to get inside of us, and much valuable nourishment is thus unwillingly forced out of our stomachs and wasted, sometimes causing diarrhœa by irritating the bowels in its endeavor to find an outlet from the body in that direction. Let me here warn you, that your feelings are no indication of the amount of damage done to your stomach, as the following interesting case will demonstrate. Some years ago, a man called Alexis St. Martin received an injury to his abdomen, which resulted in a permanent opening from the exterior of his body to his stomach, by means of which all the phenomena occurring in that organ could be observed by an outsider. Now, Dr. Beaumont has put on record, that after St. Martin had freely used ardent spirits for eight or ten days, he could perceive his stomach to be in a very unhealthy condition; the lining of it being red and spotted with small ulcers; the gastric fluids were of poor quality and mixed with a thick, ropy, gluey material, while from the little ulcers a foul matter mixed with blood, resembling that which passes from the bowels in dysentery, was given out. And yet he notes that "St. Martin complains of no pain, nor shows symptoms of general indisposition, says he feels well and

has a good appetite." Dr. Beaumont adds
"That the free use of ardent spirits, wine, beer,
or any intoxicating liquor, when continued for
some days, *invariably* produced these morbid
changes." How could such a stomach possibly
prepare and furnish to the blood suitable nourishment for a weak or diseased kidney? Stop and
consider that I have told you that no disease is
so much influenced for good or for bad by the
condition of the general system as this very one
under consideration. A good condition of the
general system simply means a thorough performance of duty on the part of each and every organ
in the body. Is it possible for the organs to do
their work satisfactorily if you do not furnish
them with wholesome food? Would a locomotive
be able to pull a train of cars if the engineer
supplied the furnace with poor coal, full of impurities? Of course not; the steam generated
would be very insufficient and we would never
hear of the lightning express. Neither can the
human being live properly without pure, wholesome nourishment; and once for all, let me tell
you that a congested stomach cannot and will
not furnish to the blood this pure nutriment, and

your kidneys will suffer along with the rest of your organs, only, being weaker, the result to them will be more disastrous. A very beautiful and interesting phenomena takes place in the lungs, by which foul and unclean blood is purified, and heat is generated to maintain the temperature of the body. Let me tell you about it. The blood in circulating throughout the body not only carries nourishment to the various component parts thereof, but it receives from them the dead and decomposed particles, whose mission has been performed, and carries them to the different organs whose duty it is to remove them from the body. Many of these particles are carried into the lungs in the shape of carbon; you also receive into your lungs a certain amount of oxygen in the air which you breathe; now, carbon has a great affinity or liking for oxygen, so when these two substances are brought into contact in the lungs, they rush together and unite, forming carbonic acid (which is a mixture of carbon and oxygen), and as such many of the dead and consequently injurious particles are removed from the body in expiration. The importance of this process will be understood when I tell you that the equivalent of about eight ounces of solid carbon is thus removed from the body in twenty-four hours. You all know how poisonous are the effects produced by the inhalation of carbonic acid gas, and therefore can readily appreciate how very injurious the retention of such a large amount of carbon in the body would be. The chemical union of carbon and oxygen generates heat, just as the burning of coal in your stove does, the carbon in the coal uniting with the oxygen in the air. This heat serves to keep up to its proper standard the temperature of the body. If its production be interfered with to any great extent, you would soon freeze to death on a cold day. Now, let us see what occurs when alcohol is circulating in the blood. Carried into the lungs, it has a greater affinity, a stronger liking for the oxygen in the inspired air than the carbon in the dead tissue has, hence it rushes forward and eagerly appropriates it to its own use, leaving the decomposed particles to remain in and poison the blood, and through it, the whole system. This great affinity which alcohol has for oxygen will also cause an excessive generation of heat, just as a good draught, by increasing the supply of oxygen to a

stove, will augment the fire and increase the heat
given off. This extra heat will render the blood
feverish and unfit to properly nourish the kidneys,
while the impurities retained in the blood will
react on the stomach, through the circulation,
and impair digestion and the proper preparation
of the food. Many of these decayed elements
which have been unable to get out of the body
through the lungs, will then be carried by the
circulation to the liver, which organ will try to
eliminate them, and for a time will succeed in
doing so ; but this extra work thrown on the liver,
in addition to its own proper functions, will
sooner or later exhaust it, and disease of this
organ will result. I will now simply tell you
that alcohol has an injurious effect on each and
every organ of the body, peculiar to the particular
organ, but will not detail them in this little book,
passing on to its specific action on the weak or
diseased kidney. I have known some physicians
to recommend the use of a small drink of whiskey
before dinner, in Bright's Disease, with a view to
stimulating the appetite and aiding digestion.
Let me tell you what this small dose of whiskey
will do, and you will clearly understand why its

use will be injurious, and should not be indulged
in. It will stimulate the appetite and, at first,
will aid digestion; so far the physicians who
advocate its use have some reason for their directions. But let us look further. The appetite
is the voice of the system, demanding nourishment
to repair its waste, and is made known through
the agency of the stomach, which organ it uses
as its mouth-piece, to make us cognizant of its
wants. Of course there is a definite amount of
nourishment needed to repair a definite amount of
waste, and all food which is taken over and above
what is required to renew the tissue used up is
unnecessary. The body not requiring it must get
rid of it in some way. If an excess of fatty food
is eaten, it will in many people be stored up in
the system in the shape of fat for future use. Let
us suppose that you have meat for dinner and
take a drink of whiskey before sitting down to
table. Your appetite is increased; a certain portion of this appetite is due to the alcohol and is
not a *natural* demand of the body for nourishment. Therefore, you will eat to excess if you
satisfy this appetite. The stomach, stimulated by
the alcohol, will digest more meat than the system

requires, the blood will take it up and after making its rounds and repairing all the waste, will have a balance remaining, which it must get rid of. Meat is rich in *nitrogen.* I have already told you that one function of the kidneys is to eliminate *nitrogenous* elements; so this excess of *nitrogenized* food is carried to the weak kidneys and a demand made upon them to remove it from the blood. So that you can readily see that the good effects of alcohol, in improving the appetite and promoting digestion, in Bright's Disease, are more than counterbalanced by the excessive amount of work which its use will throw upon the weak kidneys. In conclusion, let me tell you that Dr. Christison, a very eminent physician of Great Britain, has said that from three-fourths to four-fifths of the cases of Bright's Disease with which he had met in Edinburgh were in persons who were habitual drunkards, or who, without deserving this appellation, were in the constant habit of using ardent spirits several times in the course of the day. The cases of this disease to which I have referred as occurring in my experience at the Philadelphia Hospital were generally from the lowest and most

depraved classes of society, among whom drunk-
enness, or at least the use of alcohol, is the rule.
Dr. Black, in his lectures on Bright's Disease,
when enumerating the causes of this disease, con-
cludes his list by saying, "and, par excellence,
inordinate indulgence in alcoholic liquors." If,
then, alcohol is such a prominent cause of Bright's
Disease, does it seem inconsistent to say that its
use will hasten the degeneration of the kidney,
when it has commenced? Therefore, I will enunciate the following rule concerning its use.
Never drink alcoholic liquors as a beverage, and
only use them as medicine on rare occasions, in
small quantities, in great emergencies, and under
the advice of a competent physician.

Tobacco.—We now come to alcohol's twin
brother. As long as the world lasts, I suppose
there will always exist a controversy as to the
effects of alcohol and tobacco. Those who enjoy
and desire to use these two articles will stoutly
champion their innocence ; while, on the other
hand, those opposed to their use will denounce
them as pernicious and injurious. Occasionally,
a converted smoker or drinker, his habits changed
from pure conviction, will realize and believe in

the injurious effects of tobacco and alcohol, and will join his voice and influence with the anti-smokers and the anti-drinkers. But the rule may be laid down, with few exceptions, that once a smoker or drinker always a smoker or drinker. There is one body of men, however, the medical profession, who are united, or nearly so, in proclaiming the deleterious action of these agents. You will scarcely ever find an intelligent physician who will not admit that excessive use of either alcohol or tobacco will be more or less injurious to every one. And by excess, the medical man does not mean what the general public understands by it. A much smaller quantity will constitute excess in his judgment. Three cigars daily would seem to the ordinary individual like exemplary moderation, while to one who has given the subject much thought or study, it would constitute excess. Let me stop generalizing, and tell you bluntly, that a single puff of tobacco smoke means excess to a person whose kidneys are diseased. Now, why? Because tobacco is a foreign agent, an article whose presence is not requisite to the welfare of a single organ or part of the body, and whose use will prove injurious to

the system at large, and particularly to the kidney. Nicotine is the active principle, the essence,
so to speak, of tobacco. Now, it has been clearly
demonstrated by physiological research, that
nicotine is removed from the body through the
agency of the kidneys. Remember what I have
said about the necessity of giving rest to the diseased kidneys, and then think that by taking
nicotine into your body you are forcing the kidneys to do extra work in removing it. The habit
of inhaling tobacco smoke is particularly bad, because by this means a greater amount of nicotine
is taken into the blood through the agency of the
lungs. Still, however, you can understand how
you must inhale a great deal of this poison, when
sitting in a room in which one or more persons
may be smoking, even though you are not
smoking yourself. Hence, to live strictly as you
ought to, it would be proper for you to avoid a
room in which smoking is going on. To realize
the fact that tobacco smoke must be injurious,
you have only to recall those instances among
your acquaintances where persons unaccustomed
to tobacco have been made sick even unto vom-
iting by its use in their presence, and to remem-

ber that the large majority of smokers have to be
initiated into the art of smoking, through a series
of headaches and sick stomachs. The law of
tolerance, about which I have already told you,
is well illustrated here. Any article, no matter
how injurious it may be, if commenced in very
small quantities, may finally be used in enormous
doses with little or no apparent effect. But this
tolerance does not prove their innocence. On
the contrary, the fact that a course of training is
necessary in order that their use may be tolerated,
tends to prove that they must in themselves be
injurious; for no training is necessary in order
that the human being may use *wholesome* articles;
it is natural for him to do so. Again, a portion
of nicotine will be carried into the stomach in
the saliva or spittle which you swallow, and from
there absorbed into the blood. If you expecto-
rate all your saliva to escape this evil, you incur
an equally great one in imperfect digestion and
preparation of your food, because the presence of
saliva is necessary to the proper digestion of
some of the articles of food. Again, the excessive
use of tobacco will upset the stomach, and interfere with the thorough digestion and preparation

of your food ; so you can see that poor blood will then furnish poor nourishment to your kidneys. Let me then sum up by telling you that if you want to live *strictly* as you ought to in this disease, you will throw away all of your tobacco. If you consider this too much of a deprivation and are unwilling to do it, reduce the quantity consumed as much as possible; one cigar daily, in the evening, after a hearty meal, will work you the minimum of harm and will really constitute moderation. Remember that any use of tobacco is particularly injurious to the diseased kidney, and that the amount of injury done will be directly in proportion to the amount of tobacco consumed.

Your diet is of great importance in this disease. In the first place, because so much depends on the condition of the general system, which cannot be in good order unless a sufficient amount of nourishing food is taken into your stomach. In the second place, because certain articles of food seem to have a special determination to the kidneys. This last indication only applies to cases in which an excess of food has been taken, more than the system requires, when it becomes neces-

sary for the various eliminatory organs to remove
this excess. As I have already told you, one
function of the kidney is to remove from the body
the nitrogenous elements. Beef is very rich in
nitrogen; hence you should be careful not to eat
too much beef, or you will throw an excess of
work on your kidneys. At the same time, beef
is very nourishing, and forms muscle; hence an
insufficient quantity of it would interfere with the
proper and necessary production of vigor and
strength. So, with beef, as with all other articles
of food, you must endeavor to select the proper
medium. This you can only do by experience.
The amount of the various kinds of food required
in twenty-four hours to support life and strength
in the average man in good health and vigor has
been determined upon, but in a practical work
like this would be of no use, because every man
must really be a law unto himself in this matter
of eating. The amount required is influenced by
many circumstances. One who is employed at
hard work or takes much exercise obviously requires more nourishment than one who idles
through life. Let every one observe the following
rule. Chew all of your food thoroughly, swallow

it slowly, and cease eating before you experience a sense of fullness and discomfort in the stomach. When the chemist or druggist wishes to make a chemical solution, you all know that it is necessary for him to first put his solids in a mortar and thoroughly and finely pulverize them with a pestle before he adds his liquid which is to make the solution; if he did not, the fluid would only be able to act on and dissolve the outer layer of the solid, while the inner portions would remain intact. Digestion is also a chemical solution of your food through the agency of the liquid gastric juice. Your mouth is the mortar and your teeth the pestle. If you do not thoroughly grind and pulverize your food with your teeth, you can understand how impossible it is for the gastric juice to reach all parts of it. If you eat until you feel full and drowsy and have a sense of weight in your stomach, you can rest assured that you have eaten too much. If your stomach be strong and vigorous, it will digest all you have taken, and your poor kidneys will have to work off a portion of the excess. Frequently repeated, this over repletion will eventually exhaust and ruin the stomach, and the suffering and manifold evil

E

effects of dyspepsia will result. Use the judgment
with which the Almighty Creator has endowed
you, and, by carefully noting the effects which
the various articles of food have upon your stomachs and your feelings, you will soon be able to
make up a suitable bill of fare. The old saying,
that "at thirty years of age every man is either
his own physician or a fool," is eminently true in
regard to his diet. No *absolute* rules can be laid
down for your guidance, but experience must
regulate the quantity and quality of diet for each
individual. Let me sum up by giving you an excellent example to follow, which I quote from the
valuable letter of the Hon. Eli K. Price, already
referred to. He says "I have always been careful of my diet, but not dainty or fastidious. I
have eaten to live with a comfortable stomach and
a clear intellect, and have constantly watched
these indexes. I am as watchful as to my food as
is the smelter of iron that his furnace shall not
chill or choke, and regulate my food to prevent
constipation or laxity, rather than to resort to
medicine, which I avoid using until necessary."
The plainer your food the more wholesome will it
be. In warm weather you will require less meat

and less oily food than in cold weather. So, in a word, regulate your diet according to the results of your experience, and remember that you are more apt to suffer from *over* than from *under* eating, and exercise the greatest care not to gorge yourself. I will for an instant refer to and recommend the free use of milk, because it is very nourishing and exceedingly easy of digestion. In some persons, however, its use will cause biliousness and constipation, and here it must be proscribed. If it does not produce these or other bad effects, its very free use will prove beneficial. When the kidneys have become very much degenerated and a large amount of urea has accumulated in the blood, we often find a great diarrhœa or purging set up, and we discover urea to exist in the matter ejected. Now this shows an ability on the part of the bowels to help and assist the weak kidneys in the performance of their duty. So, when the kidneys are only slightly diseased and only a small amount of urea is left in the system, the bowels will endeavor to remove it and help their fellow organs. Now you can understand the importance of regular and full evacuations from the bowels. Again, if you make a fire in your

stove, the coal will soon burn to ashes, and if you
do not remove the ashes when you add fresh coal,
your fire will go out. The dead coal, in the
shape of ashes, will choke and finally extinguish
the flame. The contents of the bowels are the
human ashes, the result of the death and decay of
tissue, and are not only no longer of any use to
the body, but their presence will be absolutely
injurious. You must get rid of them. To any of
my readers who may be of the constipated habit,
I will say that to enjoy good health you *must*
have a *daily* evacuation from the bowels. Select
the most convenient hour in the day, and always
go to the water-closet precisely at the same time.
Regularity is a very important factor in coaxing
the torpid and languid bowels to regular action.
Combined with this regularity of habit I have, in
numerous instances, found a glass of *cold* water,
drank after sitting down to breakfast and *immediately* before eating, very efficacious. Some people
with delicate and sensitive stomachs will com-
plain that this water will make them sick. If
they persevere in its use for a few days, and take
it *immediately on the instant* before commencing
to eat, it will not make them sick. Assistance

may also be rendered to the bowels by eating fruit before breakfast, by using oatmeal, bran bread, etc. If these simple household means fail to overcome the constipation, you should at once consult some good physician. As you value your health, do not use opening medicines on your own responsibility. I venture to say that there is not a solitary physician in active practice who has not, or has not had, some patients, principally women of middle age, who, to use a vulgar but expressive phrase, are *all broken up.* A *natural* evacuation is as impossible to them as it would be to place the Sphynx of Egypt upon the dome of St. Peters, and all because they have for years been in the habit of using medicine to secure a passage. The bowels have become so accustomed to this medicinal stimulus that they positively refuse to act without it, and such cases are among the most intractable that the physician is called upon to treat, in some cases defying the most skillful management. Now, I want you all to realize the necessity of full daily evacuations, as well as the importance of securing them by natural means and without the resort to drugs.

I will conclude this little treatise by telling you

of the great importance of a proper amount of
sleep. The muscles under control of your will,
or voluntary muscles, do not require rest any more
than the involuntary muscles and organs do, but
their excessive use being followed by a sense of
fatigue, you are induced to give them rest. Every
single one of the myriad phenomena which constitute life entails the destruction of some particles of tissue which have been concerned in its
performance. The acts of thinking, of seeing, of
hearing, of smelling, of tasting, in fact, every
action of the economy, will cause the death and
decay of some particles which must be removed
from the body. You will then understand that
the kidneys are in a state of activity, are constantly working when we are awake and *actively
living*, so to speak. On the other hand, when
we are asleep many of our senses are for the time
being dormant; our whole life is lessened in in-
tensity; we live slower, as it were. Hence there
must, of course, be less destruction of tissue, in
consequence of which the kidneys, as well as all
the other eliminatory organs of the body, are enabled to secure repose and rest, during which time
they can repair the waste in themselves and re-

cupetate their strength for future work. I would say, as a rule, that every one should secure at least eight hours of *sleep* out of the twenty-four.

Put the contents of my little book into your mental stomach and thoroughly digest them. The result will be that you will understand that all I have told you really means that you should live *reasonably* and *moderately*. Herein lies the whole secret of longevity, whether you be sick or well. If I have succeeded in making clear to non-professional minds the essential nature of Bright's Disease, and the fact that one afflicted with it may live and enjoy himself for many years if he will only use common sense, prudence and experience, and if I have made him understand how to use these valuable agents, I shall have fully accomplished my desire. Hoping that I may have accomplished this purpose, even to a limited extent, I cut this little volume loose from its moorings, and trust that it may carry tidings of comfort and cheer, and prove of benefit to my friends with "Bright's Disease."

SELECT LIST OF BOOKS

FROM THE CATALOGUE OF

MR. PRESLEY BLAKISTON,

1012 Walnut Street, Philadelphia,

FOR GENERAL AND SCIENTIFIC READERS.

☞ *Any of the following books will be sent, postpaid.*

HEALTH AND HEALTHY HOMES. A Guide to Personal
and Domestic Hygiene. By George Wilson, M.A., M.D., Medical Officer
of Health. Edited by Jos. G. Richardson, Professor of Hygiene at
the University of Pennsylvania. 12mo. Cloth. 314 pp. Price $1.50.

CONTENTS.

NOTICES OF THE PRESS.

" A most useful and in every way acceptable book is that published by
Presley Blakiston, of Philadelphia, entitled ' Health and Healthy Homes ;
a Guide to Domestic Hygiene, by George Wilson, M.A., M.D., with Notes
and Additions by J. G. Richardson, M.D.' We can speak of the work of
Dr. Wilson as one of great merit and utility. It is just such a work, in fact,
as one would have expected from the author of the ' Handbook of Hygiene
and Sanitary Science,' which has reached its fourth edition. Dr. Wilson
is a Medical Officer of Health. He speaks, therefore, as a medical man
of large and special experience, and the style and structure of the book
now before us reveal the accomplished scholar, as well as the literary
adept. In the introductory chapter, in which Addison's ' Vision of Mirza'
is turned to excellent account, the author, by a skillful appeal to vital statistics, shows how vast is the amount of preventable disease and suffering.

1

Having had a good foundation, and having proved that we are, much more than we believe' we are, the custodians of our own lives and of our own health, he proceeds in a series of chapters to explain the structure of the human body and the physiology and functions of various organs, supplying all the information which is necessary to enable us to understand those intricate processes which constitute the 'miracle of life.'"—*New York Herald.*

EYESIGHT, GOOD AND BAD. The Preservation of Vision.

By Robert Brudenel Carter, M.D., F.R.C.S. With many explanatory illustrations. 12mo. Cloth. Price $1.50.

PREFACE.

A large portion of the time of every ophthalmic surgeon is occupied, day after day, in repeating to successive patients precepts and injunctions which ought to be universally known and understood. The following pages contain an endeavor to make these precepts and injunctions, and the reasons for them, plainly intelligible to those who are most concerned in their observance.

"The publications for popular use, as well as those for professional medical men and surgeons, which Mr. Presley Blakiston issues from his new establishment, 1012 Walnut street, have already won him distinction in his line. One of the latest is his edition of Professor Robert Brudenel Carter's excellent volume called "Eyesight, Good and Bad; a Treatise on the Exercise and Preservation of Vision." It makes a book of 270 pages, distributed through which are many illustrations. The nature of that most delicate organ, the eye, on which so much of the happiness of life depends, is described, along with the weaknesses it may inherit, the dangers it may be exposed to and the diseases to which it is liable as time advances. Most excellent advice for preserving it when healthy and treating it when it is impaired is given by the distinguished author, along with directions concerning the proper glasses that may be needed. The various phenomena of color, as they affect the eyes and the vision, are described in a way that will interest all readers, and the remarks relative to the treatment of the eyes of children will be found most valuable to parents, who often find fault with little ones and their vision, when they themselves are really at fault for neglecting the eyes of the little ones."—*Philadelphia Bulletin.*

WHAT TO DO FIRST in Accidents and Poisoning. By Charles W. Dulles, M.D. Illustrated. 16mo. Cloth. Price 50 cents.

PREFACE.

Whoever has seen how invaluable. in the presence of an accident, is the man or woman with a cool head, a steady hand, and some knowledge of what is best to be done, will not fail to appreciate the desirability of possessing these qualifications. To have them in an emergency one must acquire them before it arises, and it is with the hope of aiding any who wish to prepare themselves for such demands upon their own resources that the following suggestions have been put together.

THE MANAGEMENT OF CHILDREN in Health and Disease. By Mrs. Amie M. Hale, M.D. A book for mothers. 12mo. Cloth. Price 50 cents.

WHAT THE LEADING DAILY PAPER OF PHILADELPHIA SAYS OF IT.

"No better book than this, on the management of children, is to be had in such a small compass and convenient form. The chapters on 'Food and Sleep,' 'How shall Children be Dressed,' on 'Infant Digestion and Diet,' are all valuable. Those on Indigestion, especially, will give some new ideas to mothers who are accustomed to nurse their children whenever they cry, thus often giving them still more to cry about, in the way of overloaded stomachs. One subject, in particular, should be studied, as an article of religious faith, by all delicate mothers who have given their children weak lungs and tender throats to go through life with, or when babies get their consumptive tendencies from the father's side. The ounce of precaution in childhood goes further than many pounds of medicine or years of care thereafter. All scrofulous children, whether showing symptoms of lung troubles or other, should be taken in hand at once, and what is called a prophylactic treatment applied. In other words, give what food or medicines are needed to overcome these *tendencies;* do not wait until these break out, in after years, into decided symptoms. Children can learn to take cod-liver oil—if not to cry for it, at least to like it—and by taking all these agents, milk and the strengthening oils, that supply what the parents have not given by way of outfit, tone and health to the system, many a weak and apparently fore-doomed child has outgrown its dreadful inheritance and lived to a healthy old age. Begin with the children. For other and the sudden diseases of childhood, Dr. Hale's book gives wise and encouraging advice. Altogether, it is a book which ought to be put into every baby basket, even if some lace-trimmed finery is left out, and should certainly stand on every nursery bureau."—*The Philadelphia Ledger.*

BIBLE HYGIENE; or, Health Hints. By a Physician. This book has been written, first, to impart in a popular and condensed form the elements of hygiene. Second, to show how varied and important are the Health Hints contained in the Bible, and third, to prove that the secondary trendings of modern philosophy run in a parallel direction with the primary light of the Bible. 12mo. Cloth. Price $1.25.

THE AMERICAN HEALTH PRIMERS. Edited by W. W. Keen, M.D. Bound in Cloth. Price 50 cents each.

The twelve volumes, in Handsome Cloth Box, $6.00.

I. **Hearing and How to Keep It.** With illustrations. By Chas. H. Burnett, M.D., of Philadelphia, Aurist to the Presbyterian Hospital, etc.

II. **Long Life, and How to Reach It.** By J. G. Richardson, M.D., of Philadelphia, Professor of Hygiene in the University of Pennsylvania.

III. **The Summer and Its Diseases.** By James C. Wilson, M.D., of Philadelphia, Lecturer on Physical Diagnosis in Jefferson Medical College.

IV. **Eyesight, and How to Care for It.** With Illustrations. By George C. Harlan, M.D., of Philadelphia, Surgeon to the Wills (Eye) Hospital.

V. **The Throat and the Voice.** With illustrations. By J. Solis Cohen, M.D., of Philadelphia, Lecturer on Diseases of the Throat in Jefferson Medical College, etc.

VI. **The Winter and Its Dangers.** By Hamilton Osgood, M.D., of Boston, Editorial Staff Boston *Medical and Surgical Journal.*

VII. **The Mouth and the Teeth.** With illustrations. By J. W. White, M.D., D.D.S., of Philadelphia, Editor of the *Dental Cosmos.*

VIII. **Brain Work and Overwork.** By H. C. Wood, Jr., M.D., of Philadelphia, Clinical Professor of Nervous Diseases in the University of Pennsylvania, etc.

IX. **Our Homes.** With illustrations. By Henry Hartshorne, M.D., of Philadelphia, formerly Professor of Hygiene in the University of Pennsylvania.

X. **The Skin in Health and Disease.** By L. D. Bulkley, M.D., of New York, Physician to the Skin Department of the Demilt Dispensary and of the New York Hospital.

XI. Sea Air and Sea Bathing. By John H. Packard, M.D., of Philadelphia, Surgeon to the Episcopal Hospital.

XII. School and Industrial Hygiene. By D. F. Lincoln, M.D., of Boston, Mass., Chairman Department of Health, American Social Science Association.

NOTICES OF THE PRESS.

"This is volume No. 5 of the 'American Health Primers,' each of which *The Inter-Ocean* has had the pleasure to commend. In their practical teachings, learning, and sound sense, these volumes are worthy of all the compliments they have received. They teach what every man and woman should know, and yet what nine-tenths of the intelligent class are ignorant of, or at best, have but a smattering knowledge of."—*Chicago Inter-Ocean.*

"The series of American Health Primers, edited by Dr. Keen, of Philadelphia, and published by Presley Blakiston, deserves hearty commendation. These handbooks of practical suggestion are prepared by men whose professional competence is beyond question, and, for the most part, by those who have made the subject treated the specific study of their lives. Such was the little manual on 'Hearing,' compiled by a well-known aurist, and we now have a companion treatise, in *Eyesight and How to Care for It,* by Dr. George C. Harlan, surgeon to the Wills Eye Hospital. The author has contrived to make his theme intelligible and even interesting to the young by a judicious avoidance of technical language, and the occasional introduction of historical allusion. His simple and felicitous method of handling a difficult subject is conspicuous in the discussion of the diverse optical defects, both congenital and acquired, and of those injuries and diseases by which the eyesight may be impaired or lost. We are of the opinion that this little work will prove of special utility to parents and all persons intrusted with the care of the eyes."—*New York Sun.*

"The series of American Health Primers, now in course of publication, is presenting a large body of sound advice on various subjects, in a form which is at once attractive and serviceable. The several writers seem to hit the happy mean between the too technical and the too popular. They advise in a general way, without talking in such a manner as to make their readers begin to feel their own pulses, or to tinker their bodies without medical advice."—*Sunday-school Times.*

"*Brain Work and Overwork.* By Dr. H. C. Wood, Clinical Professor of Nervous diseases in the University of Pennsylvania. This is another volume of the admirable "Health Primers," published by Presley Blakiston. To city people this will prove the most valuable work of the series. It gives, in a condensed and practical form, just that information which is of such vital importance to sedentary men. It treats the whole subject of brain work and overwork, of rest, and recreation, and exercise in a plain and practical way, and yet with the authority of thorough and scientific knowledge. No man who values his health and his working power should fail to supply himself with this valuable little book."— *State Gazette, Trenton, N. J.*

ON SLIGHT AILMENTS. Their Nature and Treatment.
By Lionel S. Beale, M.D. Large 12mo. Cloth. Price $1.75.

Among civilized nations a perfectly healthy individual seems to be tho exception rather than the rule ; almost every one has experienced very frequent departures, of one kind or another, from the healthy state ; in most instances these derangements are slight, though perhaps showing very grave symptoms, needing a plain but quick remedy.

CONDENSATION OF CONTENTS.

The Tongue in Health and Slight Ailments, Appetite, Nausea, Thirst, Hunger, Indigestion, its Nature and Treatment, Dyspepsia, Constipation, and its Treatment, Diarrhœa, Vertigo, Giddiness, Biliousness, Sick Headache, Neuralgia, Rheumatism, on the Feverish and Inflammatory State, the Changes in Fever and Inflammation, Common Forms of Slight Inflammation, Nervousness, Wakefulness, Restlessness, etc., etc.

OTHER BOOKS BY DR. LIONEL S. BEALE, F.R.S., F.R.C.P.

DISEASE GERMS. Their Real and Supposed Nature and
their Destruction. 2d edition, 117 illustrations. 12mo. Cloth. Price $4.00.

BIOPLASM. A Contribution to the Physiology of Life. Illus-
trated. 12mo. Cloth. Price $2.25.

PROTOPLASM. Or Matter and Life. 3d edition. 16 Colored
Plates. 12mo. Cloth. Price $3.00.

THE MICROSCOPE. How to Work with It. A Complete
Manual of Microscopical Manipulation. 400 Illustrations. 8vo. Cloth. Price $7.50.

THE MICROSCOPE IN PRACTICAL MEDICINE. With
full directions for examining, preparing and injecting objects, the various secretions, etc. By Lionel S. Beale, M.D. 4th edition. 500 illustrations. 8vo. Cloth. Price $7.50.

THE ART OF PERFUMERY. The Methods of Obtaining
the Odors of Plants and Instruction for the Manufacture of Perfumery, Dentifrices, Soap, etc. etc. By G. W. Septimus Piesse. 4th edition enlarged. 366 illustrations. 8vo. Cloth. Price $5 50.

WORKS ON HYGIENE, CLIMATE, ETC.

SANITARY EXAMINATION OF WATER, AIR AND Food. By Cornelius B. Fox, M. D. 94 engravings. 12mo. Cloth. Price $4 00.

NUTRITION IN HEALTH AND DISEASE. A Contribution to Hygiene and Medicine. 3d edition. By J. Henry Bennett, M.D. 8vo. Cloth. Price $2.50.

HYGIENE AND CLIMATE in the Treatment of Consumption. 3d edition. By J. Henry Bennett, M.D. 8vo. Cloth. Price $2.50.

PRACTICAL HYGIENE. A Complete Manual for Army and Civil Medical Officers, Boards of Health, Engineers and Sanitarians. 5th edition. With many illustrations. By Ed. A. Parkes, M.D. 8vo. Cloth. Price $6.00.

VOCAL HYGIENE AND PHYSIOLOGY. With special reference to the Cultivation and Preservation of the Voice. For Singers and Speakers. With engravings. By Gordon Holmes, M.D. 12mo. Cloth. Price $2.00.

HEALTH RESORTS of Europe, Asia and Africa. The result of the Author's own observations during several years of health travel in many lands. By T. M. Maddçn, M.D. 8vo. Cloth. Price $2.50.

THE OCEAN AS A HEALTH RESORT. A Handbook of Practical Information as to Sea Voyages. For the Use of Invalids and Tourists. By Wm. S. Wilson, M.D. Illustrated by a chart shewing the ocean routes of steamers, and the physical geography of the sea. 8vo. Cloth. Price $2.50.

DWELLING HOUSES and Their Sanitary Arrangements and Construction. By W. H. Corfield. Illustrated. 12mo. Cloth. Price $1.25.

WATER ANALYSIS For Sanitary Purposes, with Hints for the Interpretation of Results. By E. Frankland, PH.D , D.C.L. Illustrated. 12mo. Cloth. Price $1.00.

MISCELLANEOUS.

CHEMISTRY, INORGANIC AND ORGANIC. With Experiments and a Comparison of Equivalent and Molecular Formulæ. 295 Engravings. By C. L. Bloxam. 4th London edition revised. 8vo. Cloth. Price $4.00.

A PRIMER OF CHEMISTRY. Including Analysis. By Arthur Vacher. 32mo. Cloth. Price 50 cents.

COMMERCIAL ORGANIC ANALYSIS. Being a Treatise on the Properties, Proximate Analytical Examination, and Modes of Assaying the various Organic Chemicals and Preparations employed in the Arts, Manufactures, Medicine, etc.; with Concise Methods for the Detection and Determination of their Impurities, Adulterations, and Products of Decomposition. Vol. i.—Cyanogen Compounds, Alcohols and their Derivatives, Phenols, Acids, etc. 8vo. Cloth. Price $3.50.

ON HOSPITALS AND PAYING WARDS Throughout the World. Facts in Support of a Rearrangement of the System of Medical Relief. By Henry C. Burdett. 8vo. Cloth. Price $2.25.

COTTAGE HOSPITALS; Their Origin, Progress and Management 2d edition, enlarged and illus. By Henry C. Burdett. $4.50

HOSPITAL NURSING. A Manual for all engaged in Nursing the Sick. 12mo. Cloth. Price $1.00.

DEFECTS OF SIGHT AND HEARING; Their Nature, Causes and Prevention. By T. Wharton Jones, F.R.S. 2d edition. 12mo. Cloth. Price 50 cents.

IMPERFECT DIGESTION; Its Causes and Treatment. By Arthur Leared, M.D., F.R.C.P. 6th edition. 12mo. Cloth. Price $1.50.

COMPEND OF DOMESTIC MEDICINE, and Companion to the Medicine Chest. By Savory and Moore. Illustrated. 12mo. Cloth. Price 50 cents.

THE TRAINING OF NURSES. Their Efficient Training for Hospital and Private Practice. By Wm. Robert Smith. Illustrated 12mo. Cloth. Price $2.00.

www.ingramcontent.com/pod-product-compliance
Lightning Source LLC
Chambersburg PA
CBHW022104210326
41519CB00056B/1198